You Universe

The Shape of Life

And Your Own

First Edition

*Attikus asserts the moral right
to be identified
as the author of this work.*

© Attikus 1998 – 2025

```
    M
M       M
    M
```

A Benoit

chapters

three questions

pattern

Know Thyself

history – or fractals?

time

belief & reality

natural time

things have causes

three questions

Some questions are simple.

You take a single step to answer them.

Other questions are bigger.

Answering them is a journey of many steps.

Some questions are so big, we spend our entire lives exploring them.

It's not about a perfect answer.

It's about the journey of exploration.

There are Three Big Questions that we are all asking, all the time.

Who am I?

What is life?

What is my purpose?

Our lives are an exploration of these questions.

They are closely linked.

When the answer to one of them changes, the other two change as well.

We rearrange our lives to ask these questions.

That's how important they are.

In fact the Three Big Questions are *part of us*.

They are the roots and pathways of every human life ever lived.

We explore them as nations, individuals and communities.

These questions are at the heart of philosophy.

For this reason *you are a philosopher*.

But there's more.

Three Key Senses

Your responses to the Three Big Questions give you three core pillars in life.

These are the **Three Key Senses**.

Who am I?

Your answer to this is your

Sense of Self.

What is life (or reality)?

Your answer to this is your

Sense of Life.

What is my purpose in life?

Your answer to this is your

Sense of Purpose.

We actually live out the Three Big Questions and the Three Key Senses as the voyage of life.

This is why you are even more than a philosopher.

You are philosophy itself.

Conscious Philosopher

Since you are a philosopher, you may as well be a Conscious Philosopher.

The Conscious Philosopher is *aware* of asking these questions.

She considers them deeply and *chooses* her responses.

When we are more aware of our choices, we get a better understanding of the lessons and gifts they bring.

We have more clarity around the Key Senses.

We make better decisions.

We live better lives.

We are here to choose.

pattern

*a
drop
of rain
falls on
a leaf*

*another
raindrop
falls
on the
same tree
on a low
ridge
by a lake
in Yellowstone
Wyoming*

*one
of these drops
through Idaho
and Oregon
to Pacific
flows
the other
through Missouri,
Mississippi rivers
to the Gulf of
Mexico*

The raindrops are now 2,000 miles apart.

Small things make a big difference – it was a gust of wind that made it so.

tree and river

Imagine you are an eagle flying above the river.

What do you see?

Branches, creeks and twigs.

Is this a river or a tree?

The shape of the river is the shape of the tree.

The veins of a leaf have the same shape too.

The pattern you see is a *fractal*.

Fractals are everywhere in nature – even in you.

Look at the back of your hand (or forearm).

The veins are like a river delta.

The blood moves through you in a *fractal cycle*.

It's 60,000 miles long.

It keeps you alive.

There are parts you can see and parts that you can't.

around the world
rain
drops
on
tips
of trees
trunks
and slopes
to feed

creeks
that feed
rivers
that feed

seas and rise
to
clouds
that fall
as

rain
and

snow
on
tips
of
trees

This poem is about the movement of water. It's also about the land and cycles.

All fractals have something to do with cycles.

fractal ocean

How many oceans are there in the world?

All the oceans connect as one.

Although we give it different names, like the blood in your body, this is a single system.

This *world ocean* is also a fractal cycle.

With its rivery veins and arteries, it looks a lot like you.

It's part of an even larger fractal.

The *water cycle* is all the water of the Earth moving in different forms.

Its flows are big or small, fast and slow.

The water cycle also has parts you can see and parts that you can't.

The water of the Earth may be older than the sun.

So next time you see a raindrop, ask:

How many times have you fallen as rain?

fractals in fractals

There's so much iron in the Earth — 200 billion cubic miles of it in liquid form alone (2000 miles beneath your feet).

There's a lot of iron in you too.

That's no coincidence.

Your DNA coils in a circuit.

You are a magnetic field with a body – just like the Earth.

The iron in you is part of the iron cycle of the Earth.

The same is true for other elements too.

liquid mountains

The Earth is made of fractal cycles.

You can even see spirals of water in a river or within the turning clouds.

When you look inside a fractal, there's a map of the whole thing.

Look at the side of a mountain.

Look at the side of a wave.

They are both fractals.

You can see smaller waves in both of them.

a mountain is a wave in slow motion

*mountains are waves
made of rock
moving slow
so
we know not
they flow
yet
there are things
so fast in the air
we know not
they are there*

Everything solid is a wave.

Everything solid is liquid at a slower speed.

Everything invisible is real.

Everything real builds in waves.

Everything alive breathes.

A breath is a wave.

A wave is a breath.

A wave ends as it begins. That's a cycle.

The universe is cycles within cycles.

the art of seeing

Wherever you are in the world, stop and look around.

Whatever you are looking at, ask:

Where do you come from?

Everything you can see – living or not – is from the same place: the Earth.

With every breath and heartbeat, Earth gives us life in a thousand different ways.

The Earth is not just dirt and rocks.

This is a living system that includes you and me.

Our bodies are the Earth.

What we do to the Earth we do to ourselves.

What we do *for* the Earth we do *for* ourselves.

a map of everything

Fractals are part of larger fractals – sometimes much larger.

The electrons in the atoms of your body look like planets spinning around their stars.

The elements within us are in the stars of the Milky Way.

The planet Mercury has a crust of diamonds nine miles deep.

Twinkle, twinkle, little planet.

One thing becomes another.

Every year you replace almost all the atoms of your body.

Everything used to be something else.

With each breath your body is new.

What is you today will be the world tomorrow.

The same is true for other stars in the sky.

In a way you are even part of the sun.

If you drew a map of where everything comes from and where everything goes to, it would look like a cosmic river, a tree of life unfolding.

A fractal map would show it all.

ancestor tree

How many ancestors do you have?

One generation back, you have two parents.

Two generations back, you have four grandparents.

Three generations back, you have eight great-grandparents.

20 generations back, you have a million ancestors.

That's a million people who made you, *you*.

If you drew a map of your ancestors, what would you see?

There's a reason it's called a *family tree*: ancestry is fractal.

It's part of a larger fractal too.

All our ancestries join to form a single human family tree.

Everyone you know is in this *fractality*.

It includes our stories and histories.

All those yet to be born will join the river of ancestry.

Our human family is a branch of a still larger tree.

from the smallest twig
to the Mother Tree
the forest grows
fractality

evolution

We study different species and how they evolve.

In this we tend to think of species as separate from each other.

There's another way of looking at it.

Life responds to various conditions.

Each species is a different response *by life*.

When you map all species together, it looks like a tree.

Any group of species within this also resembles a tree.

Evolution is fractal.

evolution now

We are all actors in a timeless tale.

It doesn't begin or end with the Earth.

The biggest fractal you can imagine is part of something vaster.

All species alive today are part of life still evolving.

Slower processes are invisible to us – but they are still happening.

Think of a growing tree.

Sometimes we might be looking at the same thing happening at different speeds but giving it different names.

A species evolves over thousands of years.

That's life adapting to conditions.

The networks of your brain create a thought.

That's life adapting to conditions.

Each thought you have is a new species.

Each life on Earth is a new universe.

The thought you are having right now is evolution in real time.

You have a front row seat on evolution: you are the universe regarding itself.

Welcome home.

universal nature

The universe doesn't start at the atmosphere.

It doesn't stop at your skin.

Everywhere in the universe *is* the universe.

That includes you.

Nature is not outside you.

That's why we need nature in our lives.

Without it we are lonely and forget who we are.

The same nature is you, the Earth, moon and stars.

trees and universes

Human evolution is entwined with language and culture.

The way these evolve also looks like a tree.

Language and culture are powered by the same thing as evolution.

Fractals allow life to infinitely explore and expand.

Just ask a tree – or a forest?

In Utah there's a trembling aspen that covers 106 acres. It's a single organism.

Trees are interesting fractals.

Each one adapts to local conditions, just like you do.

Each species has a different way of exploring space and sunlight.

The tree is a map of our lives at a different speed.

Our roots are into the Earth. We even have crowns!

A forest is the universe.

In a forest, every twig, branch and root together form a single fractal.

This includes a vast underground network that feeds the entire community.

Mycelium is a universe in itself.

source code

Is there a source code for life?

A path branches into two or more new paths.

*Where this happens is a **fractal deciding point**.*

Each path is a process, journey or outcome.

It leads to the next deciding point.

With this simple pattern, fractals map life's journeys.

It could fill universes.

It could go on forever.

This is a map of the universe that we can't see.

It works at all scales.

The bigger journeys have the same shape as the smaller ones.

The universe is a fractal made of fractals.

All fractals that we can and cannot imagine form a singular **fractal of fractals**:

Life and consciousness exploring, expanding and evolving.

The universe includes consciousness.

It's whole.

Know Thyself

The universe is a hologram.

Within a hologram, each part contains the whole.

This means you can see the entire hologram within any part of it.

Sounds like a fractal, doesn't it?

You are a fractal of the universe.

Thus you are the universe.

There are clues about you in the universe.

There are clues about the universe in you.

The universe began as a tiny dot.

This unfolded rapidly then began to slow.

A human life is the same: things change fast but then slow down.

A human life starts as a single cell.

This divides in two.

Each of these divides again.

And so it goes.

This is the universal fractal unfolding in you.

When there was *no thing*, the early universe birthed the foundations for existence – and the building blocks for all fractals.

Hot exists because of cold.

Cold exists because of hot.

Pairs of opposites were born.

Between each pair, you now have a *spectrum of possibility*.

There are billions of temperatures.

There are just as many colors.

These are two spectrums of possibility.

How many might there be?

Each spectrum allows infinite journeys of life to exist.

then contrast

From this Great Variety of possibilities, **contrast** is born.

Without contrast, everything would be the same.

We need contrast to see.

We need contrast to notice anything.

a
candle
shines
most bright at night
because of contrast
between dark and light

The reason we notice change is contrast.

Change is the contrast between what was and what is.

Without contrast there would be no perception.

Everything would be and look the same.

All perceptions and perspectives exist because of contrast.

Your perspective is the contrast between you and what you are looking at.

When you stand somewhere else, you have a different perspective.

When you change, your perspective changes.

*

All these developments provided the fractal architecture for journeys of life in this universe.

As lightning from a thundercloud, the Great Variety births all fractals.

This immense structure is the *universal kitchen* from which all possibilities unfold.

This occurs in less than a moment before time.

We ourselves come into existence within this unfolding cosmic blanket.

It all tumbles out from the birth of opposites.

That is the seed moment of this fractal universe.

We are here to resolve these opposites.

The path is the middle way.

One day opposites will have served their purpose.

The universe will fold back up to where it began.

Yet endings are beginnings.

an unprecedented miracle

Your body is made of light and energy.

These are slowed down in a specific ratio.

Einstein showed us this.

The cosmic energy in one cell of your body, released, would shine brighter than the sun.

There are over 50 trillion cells in your body.

You are a miracle of the cosmos.

Did you know?

That's just your physical body.

There's nothing ordinary about you.

As a hologram, the universe can only mirror and reflect.

This is because within a hologram, everything is also everything else.

This makes the universe the perfect self-learning school for life and consciousness to evolve.

The entire universe – its nature and chosen destiny – is within you.

The door opens from your side.

history – or fractals?

*a drop
of rain
touches
a pond*

*what
was
not
becomes
so*

*possible
or impossible
now
it is
real*

*ripples
move out
upon time*

*the
water, life*

*its
surface, reality*

*a fractal
exists
that did not
before*

*it
remains
until
complete*

*this
is
the universe
too*

how did today become today?

Reality is not measured in time.

Reality is measured in journeys.

Time is something else.

Remember those two drops of rain at the start?

Fractals start small.

Butterfly wings can cause a tempest.

But the biggest fractal storm returns to the same quiet moment in the universe.

Upstream causes have downstream effects.

Fractals continue as they began.

A dot spirals outwards into wider circles.

The DNA of the fractal moves into more and more form.

The seed contains the code that began it.

A moment becomes a paradigm.

This is the essence of a fractal unfolding.

New layers of complexity show up.

Cause and effect patterns spiral open, sometimes over millions of years.

Yet in one instant a paradigm can become nothing.

the lake of existence

While something is part of the present *it is not the past*.

What we call *the present* is a series of overlapping fractals.

These all began at different moments of the past.

Each remains in place until its gifts have been fully unwrapped.

Only then does a fractal truly become past.

Until then it is a ripple on the lake of existence, overlapping other ripples.

Those also began before today.

It's hard to see a fractal from the inside.

They have their own rules and realities.

If you are born inside a fractal, you may not know.

We all live downstream of fractals that we don't understand.

the Sokrates deciding point

2,400 years ago, it was early days for democracy.

At the Battle of Marathon, 200 free Athenians had defeated thousands of Persians.

But 90 years later it was Athens that was becoming a tyrant – to other Greek states.

Now Sokrates, the free thinker, was seen as a thorn in the side of men who wanted to control power and opinion in Athens.

The Assembly was called and voted.

Sokrates was given a choice: exile or death.

Sokrates drank poison hemlock.

Evidence has even emerged that the vote was rigged.

With this event, abuse of power wrapped itself around democracy.

Democracy has yet to recover.

with the tree goes the seed

Over thousands of years, western states tumbled out of that Athenian fractal.

From them, relationships emerged between people and state that have now become global.

Much of the world lives downstream of the Sokrates deciding point.

the power of democracy

The nation belongs to the people.

In its pure and early thought, this is the undiluted power of democracy.

Freedom ran in the veins of the Athenians at Marathon.

That was the source of their miracle.

But with the death of Sokrates, a false democracy was established.

You are free to think your own thoughts – up to a point, said Athens.

The state placed the threat of violence above the mind and rights of the individual.

It set a pattern for a small number of people to veto true democracy and the human spirit.

Sokrates is more than someone who died for his beliefs.

He stands for speaking truth to power.

He stands for thinking out loud.

He stands for the nobility of life and consciousness.

He stands for the right of We, The People to build our nations from the best of ourselves.

The purpose of democracy is to protect the people from the tyranny of the few.

Only if this is secured do we have true democracy.

If it does not protect the vulnerable and the good, it is democracy in name only.

Democracy is not a competition.

It is the intelligence of the people organizing itself and evolving.

There is a reset code for democracy.

You just read it.

True democracy is fractal.

For a short while in Athens, the people *were* the nation.

That's why it's called democracy – *the rule of the people*.

Let's review the Sokrates deciding point and learn the lessons.

There are other reset codes.

For Medicine it is:

First, Do No Harm.

Never let self-interest compromise the integrity of scientific endeavor and service to humanity.

inheriting echoes

Accents are inherited and hard to shake.

These aren't just in the way we speak.

They are also in the way we think.

Language weaves history and culture.

Words help us track the tides of time.

There have been many versions of Rome.

Caesar became emperor then *kaiser* and *tsar*.

Our world is full of Roman concrete.

Our year is full of Roman months.

There's a saying that *history repeats itself*.

History doesn't repeat itself.

Fractals remain open until complete.

History doesn't repeat itself.

Fractals echo like voices in a canyon, until we learn from them.

Then they disappear.

There are other fractals from history that aren't yet history.

The key to the future is often unlocking the past.

This may be because to the parts of us that experience life most deeply, there is no time.

master reset code

There's one reset code that contains all others.

It's relevant to our personal journeys as well.

Sokrates found it.

I know nothing.

(Or *all I know is that I know nothing.*)

This is where East meets West.

It's also where South meets North.

Why is this statement so powerful?

How much of everything that there is to know do you know?

1%?

Less than 1%?

The vast majority of life is unknown to us.

We will discover things but this will always remain the case.

When we accept that *we know nothing*, we align with this core truth of life.

Until then we are defending ignorance, clothed in pretense.

Not knowing isn't the source of anxiety.

It's a place of peace, a kind of powerful medicine that reveals whatever it needs to reveal.

Sometimes it takes courage to say, *I don't know anything*, or just, *I don't know*.

Yet this is where all learning and all science begin.

It's the birth of self-awareness, all universes and all paradigms.

It's where we become something new.

It's also the completion point.

take

some

time

history the movie

You walk into a room.

In front of you a movie is playing.

The film is paused on a single image.

You haven't seen the movie before.

All you can see is the frame in front of you.

Everyone you know was born within this image.

The name we give the image is *the present*.

It's all you have known of the entire movie.

There were thousands of images before it.

The name we give those is *history*.

there were many presents

No one knows the story.

No one alive today has direct knowledge of the past.

We don't truly know how the past became the present.

We just have anecdotes.

There was more than one past.

Each of them was once the present.

Altogether they explain how today became today.

If you could watch the entire movie, you would understand the cause and rich background of everything we see in the world today.

each present was
as real to those
alive then
as today is
to you
or me

Seeing life in fractals makes the past come alive.

It can do the same for the future.

Fractals show us how one thing becomes another.

We can use that knowledge to understand where we are and where we want to go from here.

today began before today

Wherever you are in the world, stop and look around.

Can you see anything that began in the present?

Everything you are looking at began in the past – even us.

Our bodies began in the past.

They are still present.

What we call the present is a snapshot of fractals that began at different moments of the past.

Each is with us until it isn't here anymore.

This means that there is a very big question to answer.

If the present is made of things that are not actually past:

What is time?

time

人

Our

time

beliefs

are

creating

the

world

人

time lines

The version of time that we have come to believe in is a straight line.

This line begins in the past, moves through a blip called the present and then into the future.

We can call this *Straight Line Time*.

Straight Line Time doesn't exist.

In fact we made it up.

Do you believe me?

Let's go and stand at the South Pole.

What time is it?

It's every second of the day and night.

Every time zone converges here.

You are standing in one of two places in the world that has escaped the fiction of Straight Line Time.

The other is the North Pole.

The lines that we use for Straight Line Time are man-made.

They aren't there.

Go to space.

You won't find them.

Only a thin layer of atmosphere.

Now let's travel to a small island in the Pacific.

Stand in the right place and jump from one foot to the other.

Time traveler!

You are hopping back and forth across the International Date Line.

Where did this all begin?

人

the origin of time

One day in the past some men walked into a building.

They stood around a globe and drew lines on it.

Things have been a bit odd ever since.

Then they used those lines to come up with another idea.

They decided to call the spaces between the lines *time zones*.

This is how the version of time that you and I believe is real came into existence.

It's also why the world today looks the way it does.

人

Maybe no one ever told you the words *time is in a straight line*.

They didn't have to.

When we are young, a lot of the things we learn we simply absorb from those around us.

We learnt this model of time when we were children from people who had learnt it when they had been children.

If you keep going back in this way, you eventually find those men in the room.

We took something round and made it flat.

We took something real and turned it into an idea.

We took geography and turned it into *time*.

No wonder we are confused.

The world today is quite manic.

Are we looking for something that doesn't exist?

人

we don't live on a concept

If we assume Straight Line Time is real, there's a problem.

Straight Line Time is a concept.

We live on a planet, not a concept.

Straight Line Time is something we invented to organize stuff.

It's a tool to mark cycles and parts of cycles.

We forgot that.

We also forgot that it doesn't exist.

Then we started bowing down to it.

This has happened many times in history: inventing something then bowing down to it.

It's as if we have been looking for something to obey.

This has got us into a lot of trouble.

We should stop doing it.

making reality disappear

Straight Line Time is useful for scheduling things.

It helps us get more done.

But it's an idea, not reality.

This creates problems for us every day.

Pollution – in all its forms – only exists because of Straight Line Time.

How come?

When we use Straight Line Time, we convince ourselves that the past disappears.

人

If we *think* we are moving along a straight line from the past into the future, *in our minds* the past simply vanishes.

It stops existing.

So does anything that happened or began in the past.

This includes pollution.

That's not right.

That's not how reality works.

We live in a universe of cause and effect.

Cause creates effect.

The cause happens first.

The effect happens after.

Sometimes the effect is still happening millions of years after the cause that began it.

Straight Line Time is convincing us that causes simply evaporate.

This mistake is the basis for pollution – and many other problems that we now have on our hands.

Our confusion is staggering.

We are acting as if we are living on a different planet in a different universe.

This unconscious error applies to all sorts of things that we have set in motion before today.

These include events in our family lives and nations.

Straight Line Time is disconnecting us from the reality of cause and effect.

But cause and effect is the fundamental nature of reality in the universe.

Life is in cycles.

We live within a bio*sphere*.

When we remember these truths, we know that what we discard today we will meet tomorrow.

Pollution is a form of denial.

But it only exists because of an even bigger form of denial: Straight Line Time.

This is the power of belief.

This is the power of our beliefs about time.

one straight line, so many problems

The problems that we have caused ourselves are now so urgent that our survival depends on solving them.

So why aren't we doing that?

The reason we are not yet solving these problems is that they began in a past we have decided no longer exists.

Straight Line Time is making the causes of problems invisible.

That's why we cannot see them!

That's why we cannot solve them!

time check

We are convincing ourselves that reality is not reality.

Reality is less convinced.

We aren't doing this on purpose.

We have been doing it without realizing it.

That's why it might seem strange to think about.

Let's fix this.

Reality is waiting.

It's time.

人

the Reality of Cycles

Reality isn't in straight lines.

The world is round, not flat.

your body is the Earth

we are sitting in
a pool of things
we have released
into the
ocean, air and land

like the Earth
we absorb

~ p ~

~ ~ ~

Everything we do and everything we are part of is either a cycle or part of a cycle – even breathing.

We, the Planet, and all that is beyond the Earth, exist because of cycles, relationships and systems.

We can call this **The Reality of Cycles**.

It is the core reality of who we are and everything that gives us life.

Overusing Straight Line Time closes our eyes to all of this.

That's why we have been trashing our own home.

We only protect what we respect.

reality chasm

There's a reality chasm we need to know about.

It is between the Reality of Cycles and the *concept* of Straight Line Time.

The more that we have embedded this chasm into our thinking, beliefs and institutions, the more we have come to threaten our own existence.

人

*we live
in a real world
of cycles
spirals
waves
pulses and
rhythms*

within

*a real
universe of
cycles
spirals
waves
pulses and
rhythms*

To consider means to be *with the stars* (*con sidera*).

From your breathing lungs to the epochal rise and fall of star nurseries, life is a cycle, a wave within a wave.

It's all complex.

It's all interrelated.

Your body is complex.

The mind is complex.

Community is complex.

Ecosystems are complex.

There are no straight lines in nature.

There are no straight lines in you.

Straight Line Time denies then trashes anything that's not a straight line.

So basically everything.

人

manic distress

Why do so many of us live hurried lives?

What are we chasing?

Once we had confused Straight Line Time with reality, we started looking for more of it.

You can't find what doesn't exist.

You can't find straight lines in a circle.

There are no units of Straight Line Time to be found within the cycle of a day, month or year.

Chasing something that doesn't exist is painful.

If you believe you are running out of time, you are.

If you believe you are running out of time, how can you have enough of anything?

Greed is also rooted in the *never enough* mindset.

Code creates pattern.

The pattern expresses the code that created it.

Life is always showing us the power of our beliefs to shape our world.

The universe is helping us learn.

Are we listening?

many presents

We have come to believe that being present is not productive.

This can make being authentic seem unsafe or selfish – like a guilty pleasure.

It's not.

The present is where real experience takes place.

Being present is being human.

Being present brings quality relationships.

Being present means experiencing life in all its richness.

Being present is an act of self-balancing.

Overusing Straight Line Time is making us abandon ourselves and each other.

It's making us less human.

I just don't have the time.

How often have you heard this?

Chasing time gives us less of it, not more.

This is a modern mania.

We are throwing ourselves away in pursuit of something that doesn't exist.

we are throwing away
balance
community
wisdom
intelligence
relationships and
our ability to listen

We are throwing away our world.

What are we getting in exchange?

We have been programming ourselves for mental and physical illness.

The manic distress this causes is a psychological disorder that now affects billions of people.

It separates us from authentic experiences and relationships that are only a moment away.

When you next hear someone say *time*, ask what other words could replace it.

Attention?

Presence?

A willingness to listen?

人

why this matters

Our children are watching.

They are either copying us or despairing in a terrible realization.

They know that we have forgotten but they don't know how to reach us.

Consequently they are unprotected.

We are telling them that this very dangerous world that we are creating is what they must adapt to, put themselves aside for, stop being whole human beings for.

When we talk to them, the name we give to this educated insanity is *growing up* or *life*.

We live in the most abundant era of history.

Yet we are depriving ourselves of the basics we all need to live balanced and healthy lives.

This is sleepwalking.

It is the unexamined life.

Young people, not adults, are on the front line.

We are throwing them away every day.

They know that often we are not actually here.

What choice do they have except to shut down or lose hope?

every moment
of what we call
past or future
on all worlds
in all galaxies
and universes
is the present

There's no such thing as straight line reality.

If we think there is, we aren't living within the reality of this sumptuous planet.

Take some time to think about what I am saying.

belief & reality

人

the engine room

What we believe about time affects us in other ways.

These are to do with the way that belief works.

There's a place within us that we can call the **engine room of belief**.

Everything in here shapes our lives.

Here we find ideas and beliefs about self and life that have slipped under the radar of conscious awareness.

There's a lot of emotion too.

Everything in the engine room is being used to answer the Three Big Questions:

Who am I?

What is life?

What is my purpose in life?

Thus everything in the engine room also helps to shape *Sense of Self*, *Sense of Life* and *Sense of Purpose*.

That's how important it is for us to know ourselves.

The Conscious Philosopher finds this engine room and spends more and more time here.

She unpacks what she discovers and gives it quality attention.

This makes everything more conscious.

All these journeys benefit from the extra daylight.

This means it's no longer *stuff* directing your life.

It's you.

This engine room is where we find our beliefs about time.

人

life reflects: us

We don't realize how powerful our minds are – or how they shape our world.

In many ways, reality is the mirror to our beliefs.

Your brain reads your beliefs as instructions about what is real and what is not.

If you firmly believe something is true, your brain assumes that it's a fixed point and arranges everything else to fit in with this *Belief As Truth*.

(If you believe something is truly impossible, it could be right in front of you but you wouldn't be able to see it.)

This is how our time beliefs are shaping the world.

It also holds the key for creating a better one.

人

reality = beliefs about reality

The world we think we are living in is a mirror.

The universe is the same mirror.

This mirror reflects the power of our choices, assumptions and beliefs.

This happens whether or not we know that it's happening.

We experience reality without realizing:

Life is a canvas.

Upon this we paint.

A hologram can only reflect.

In this hologram universe, life is not showing us that our beliefs are accurate.

Life is showing us what reality looks like when we operate from a specific set of beliefs.

We are living in constant biofeedback with life.

If you don't believe me, all you have to do is change your beliefs.

人

world stories

When we have thoughts, feelings and beliefs that we don't know about, we live them out as drama in life.

This creates a world that looks like paintings of the unconscious.

It's as if we are obeying stories that we don't know belong to us.

We think we are listening to life.

We are speaking to it.

We think we are observing life.

We are instructing it which reality to present to us.

Life is responding.

The outer world is the screen onto which we project our inner questions, journeys and stories.

Do we know we are doing this?

beliefs as filters

Beliefs are filters that we use to screen reality.

They change our experience of life every day.

*without knowing it
we are asking our beliefs
what life is
how to behave and
how to make decisions*

We see as our beliefs allow us to see.

We interpret as our beliefs allow us to interpret.

But we are the ones creating our beliefs!

(There's usually a time lag between creating beliefs and those beliefs telling us what reality is.)

As a note, softening beliefs is sometimes wiser than torching them.

人

Personal Reality System

Your brain uses your beliefs to shape a personal version of reality.

This is a model of self, life and world that is unique to you.

This Personal Reality System develops over many years of layered experience.

It's based on what you have experienced so far and your responses to those experiences.

The things that have seemed most fixed, reliable or normal become the basis for your beliefs and expectations.

Everything else is considered by the brain to be some combination of irrelevant, impossible and fluid.

This is true for beliefs we recognize and those that we don't.

It's part of how reality works for us as human beings.

Straight Line Time seems to be reality because we have convinced ourselves that it is so.

Also we don't know the difference between reality and time.

人

how beliefs get stronger

Beliefs tend to reinforce themselves, especially when we don't examine them.

This applies to our time beliefs as well.

How does this happen?

When we believe something, we see evidence for it in the world *just because we believe it.*

Then we assume that our beliefs have been confirmed as truth.

That's not what's happening.

What's actually happening is that *as we go through life, we assume that we are looking through a window onto the world. In fact we are looking in a mirror.*

The universe is an extremely potent reflector.

Life shows us reflections every single day.

We receive this feedback in the form of experience and perception.

The great art is to interpret these reflections correctly.

There's a language ready to be learnt.

This is a practice that includes many things.

It involves emotions, self-awareness and self-discovery.

When we react to others we react to ourselves.

When we judge others we judge ourselves.

When we celebrate others we celebrate ourselves.

This happens through the electrons of the universe.

There are quite a few of those.

人

the innocent mind

When we take a look under the hood, we can all find beliefs we have that we didn't know we had.

There's a reason that we sometimes hypnotize ourselves with belief:

the mind is innocent.

The mind doesn't *know* reality.

The mind *assumes* reality.

We all live on faith, all the time.

That faith is made of beliefs.

We start out in life without beliefs.

The first beliefs we acquire are not because of some rational process.

They are because of personal experience.

We adapt.

As children we learn to believe things before we know what they mean – or even what belief is.

We just experience life in an absolute way.

In the early years, experience and reality are the same thing.

It's only later that we create new experiences in order to *explore* what life and reality might be.

This is why for many of us reality is still defined as our beliefs about reality.

Rivers flow from upstream to downstream.

Reality flows from cause to effect.

We can see our beliefs in the world they shape.

人

the hum of reinforcement

When you spend time with people who have similar beliefs, those beliefs often get stronger.

There's a risk that life becomes a conversation between similar people agreeing that *their* beliefs are the correct ones.

Pretty soon there's also the risk of defending beliefs that we don't actually believe.

That's called war.

This happens because those beliefs are all we have ever known, because we haven't questioned them and because they are part of our social relationships.

Such conflict is a fight against the nature of reality itself.

Our evolution in this age is to climb out of these cages of warring belief.

How?

人

We have more in common than we think.

We are all shaped by experience, loyalty and belonging.

Each of us has inherited beliefs from somewhere.

When we seem to be in conflict with each other, we are often just being loyal to different things *in similar ways*.

Perspective creates reality.

Your perspective is yours.

It depends on you.

When you change, what you are looking at seems to be different.

The universe is all perspectives and none.

Perhaps we leave the warring cages behind by understanding the feedback loop between belief and reality.

Perhaps it's by loosening beliefs before they become like granite.

Perhaps it's by choosing to be more curious, present and open.

Perhaps it's by reviewing our beliefs.

Which of them serve our shared humanity?

These will often improve our own lives.

On a bigger scale, perhaps it's the shift from *Belief As Truth* to *Belief As Perspective*.

The innocent mind is still there within us.

It's the one we all started out with.

The innocent mind would never take a precious human life because of a belief.

人

realer eyes

As we soften rigid beliefs, we see the way in which reality is constructed.

Each set of beliefs is a story – it is *mythos*.

To see how we write and live out these stories is to see the code within the code of human life.

The universe is our teacher, ready for us to *real-eyes* Who We Are.

We are here to choose and learn.

When we stop relying on Straight Line Time, we will solve the problems that it has created.

We live in a

benevolent universe.

None of us has glimpsed

this benevolence.

natural time

人

how we grow

You are not a straight line.

You are more like a tree.

Like trees, we stretch upwards in multiple directions at once.

The past gave birth to the present.

Fed by the roots of experience, who you were gave birth to who you are.

That's a fractal from a fractal.

The next time you go to see a tree, put both hands on the trunk.

How much do trees have to teach us?

This tree contains all the years of its life so far.

The earliest parts of the tree are still there.

They didn't go away.

They are the deepest parts of the *inner tree*.

We grow in the same way.

When you turned 11, you didn't stop being 10.

You became *also 11*.

This is true for all the years of your life so far.

You are not a number.

You are a fractal tree of fractal humanity.

Everything about a fractal is whole.

The child part of you is still there.

In fact your *inner child* has been on the planet longer than *you* have.

You may have learnt to leave the present.

Maybe it was unsafe for you to be your whole self.

You are still there.

Nothing good is lost.

The best things of the past, remembered or forgotten, are yours *now*.

They have been safekept for you, to be discovered anew.

Everything can be accessed when you free yourself of time *as you think you know it*.

You already do this in timeless moments of your day.

You are a unique being of the fractal forest.

Trees take care of their young and elders.

The forest takes care of the whole.

the forest is one tree

roots
      where things begin

paths
      life's journeying

time
      is in layers

memory
> is the body of the tree

change
> is constant,
> on many levels

cycles
> day and night,
> Spring to Autumn

listening
> the tree responds
> because it's listening

growing
> the tree grows
> for all its life

forest
> the tree is not alone,
> it works with others

人

buried knowing

Deep down we know that we are the same as the Earth.

This is not surface knowing.

Perhaps this is why the distortion of Straight Line Time has caused us so much trouble.

Outer mirrors inner.

What we do to the Earth we do to ourselves.

The solution to many of our problems may be simpler than *we think.*

If we use a different version of time, we will wake up as different people living in a different world.

人

why not use it?

The world we create is downstream of our beliefs.

What we believe about time shapes our world.

So why not use it?

Why not use this powerful engine to create a better world?

   if
 time
 is not
   in a
 straight
   line
  then
 what
 is its
shape
  ?

人

*every day
the sun rose
and set*

*every
one of us
breathed in
and out*

*our eyes
opened
and closed*

*our hearts
pumped
blood in a
circular
rhythm*

*the same
will be true
for every
human life yet
to be born*

人

We should choose a model of time that is aligned with the Earth, our bodies and the cosmos.

Who we are is the Reality of Cycles.

Time is in a circle.

If you have gone down the wrong path,

turn back,

no matter how far you have traveled.

~ Turkish proverb

人

the return to natural time

The solution to a problem is to address the root cause.

If we want to change, we can.

The Reality of Cycles is not just about how we view time.

The Reality of Cycles is the reality of life.

It's also the reality of fractals.

What would it be like to live in a world that looked a lot like this one?

What world do most of us want to live in?

Let's decide that.

Then let's ask,
Which model of time will get us there?

From the Reality Of Cycles comes **Circle Time.**

人

if time is

If time is in a circle, there's no rush.

It's all happening *now*.

There's no Panic Society.

If time is in a circle, you have time to be present.

You have time to be human.

You can look up from the treadmill.

If time is in a circle, causes are no longer mysteries.

We know what's important to us.

We make better decisions.

We live better lives.

If time is in a circle, then so are you.

人

Circle Time

We already know that time is in a circle. A year is a circle.

A day or a month are also circles.

The return to Circle Time is fairly simple.

We stop chasing what isn't there.

Nature is resilient.

Ecosystems are resilient.

When we remove the cause of stress, an ecosystem can restore itself.

This is because nature is a fractal hologram: the parts contain the whole.

Humanity is an ecosystem of the Earth.

It's all one fractal.

Humanity knows how to restore itself.

As we remember who we are, we will find ourselves part of a single paradigm that unites all cycles of life, within and without, to the balancing power and majesty of this Earth and universe.

人

Evolution doesn't happen on its own.

The reason we evolve is not because *time passes*.

We evolve because we do the work.

We never left wholeness.

We dreamed and forgot.

The cure is to remember the truth.

We are born as whole human beings.

Body, heart and mind form this wholeness.

But our sense of this was lost within The Forgetting – that Straight Line Time is only a concept.

Deep down we know who we are.

We are meant to thrive.

You are more than a part of the Earth.

You are to the Earth as the Earth is to the universe.

That's a circle.

I believe that our long quest has been to remember the Reality of Cycles.

Many indigenous people, including Africans, would smile to hear that.

They have never departed from the Reality of Cycles – despite violent attempts to force them to do so.

Basing our system of time on the Reality of Cycles will reconnect us as fractals of the Earth, true inhabitants of the biosphere.

Circle Time will break the spell of false time.

It will allow us to take our rightful place as balanced human beings at peace with ourselves and life.

Circle Time will align us with the stars and cosmic order of which we are made.

It will end much of the distress that we cause ourselves.

Many psychological problems will disappear.

Community will restore itself.

Biodiversity will bounce back.

Covenants between heart and mind, society and ancestry, will be reforged.

We will remember that we are all indigenous human beings.

Some people will only believe this is possible once it has begun.

It has begun.

The inner child of humanity is an indigenous girl.

Our millions of shared ancestors need us to recognize what happened. We are all this tree.

人

fractal keys

Fractals are the key.

They unlock the doors of epiphany: nature and cosmos are you and me.

The more we allow nature to restore herself, the more we are restored in kind.

A more living world will reflect to us what we can be.

when we no longer
weaponize life or
traumatize ourselves
we will unlock our potential
within this perfect fractality

Then we will resonate with the complex harmony of the stars consistent throughout the universe.

As a snowflake self-forms, Gaia – the Earth – will restore herself and us through ever richer fractal layers.

The return to Circle Time will transport us to another planet.

It will allow us to travel the stars in new ways.

The hologram universe will come to our shores.

We have nothing to fear.

Circle Time – we may even call it fractal time.

next

人

The next book in this series contains a method.

This practical method helps you answer questions that can't be answered.

It allows you to solve problems that can't be solved.

This method is aligned with the purpose of life in this universe and your part within it.

For details about
this next book,
the You Universe audiobook,
translations and
upcoming events,

email:

attikusuniverse@gmail.com

Include LIST in the subject line.

人

We are not consumers.

We are living beings of a living world, children of Mother Earth.

We rely on Earth the Mother for everything, with every breath, as our ancestors did and our descendants shall.

We are not here to consume and waste the body of the Earth, nor the generosity of life.

We Are Here
so that
at the very least
when we leave
our temporary lives
may this very gracious
majestic world
be in better shape
than when we arrived

We are not consumers.

*Any sustainable civilization protects children
the mother
families
community
and Earth*

All other paths are dead ends.

All roads lead to philosophy.

All roads are philosophia.

epithet

We only see what we *can* see and what we believe possible.

Many fractals cannot be seen.

These include the fractals of our lives.

Our paths bend beyond the corners of the known – the same curved lines as the universe?

We know that light bends.

> we know space's curved
> —
> maybe it is *even folded*

The universe is magnetic.

Printed in Dunstable, United Kingdom